IMAGES
of America

KNOXVILLE ZOO

On the Cover: Visitors stream down the hillside toward the elephant enclosure in the early 1970s. (Courtesy of Knoxville Zoological Gardens Archives.)

IMAGES
of America

KNOXVILLE ZOO

Sonya A. Haskins
Foreword by Jack Hanna

ARCADIA
PUBLISHING

ISBN 978-1-5316-2689-1

Published by Arcadia Publishing
Charleston, South Carolina

Library of Congress Catalog Card Number: 2006936542

For all general information contact Arcadia Publishing at:
Telephone 843-853-2070
Fax 843-853-0044
E-mail sales@arcadiapublishing.com
For customer service and orders:
Toll-Free 1-888-313-2665

Visit us on the Internet at www.arcadiapublishing.com

*Affectionately dedicated to my oldest child, Sarah Elizabeth,
who loves all the creatures God gave us on the fifth and sixth days.*

Every effort has been made to identify the photographers who captured the poignant and wonderful images in this book. A photograph credit is always given following each image with an identified photographer. If only the media source is known, then credit is given to the media source. Other than a very few exceptions, all the images contained within this book are the property of the Knoxville Zoological Gardens and have been taken from their archives. Many of the images belonging to the Knoxville Zoo were taken by Betty Wasserman, and I would like to thank her for helping preserve the history as well as the beauty of the animals located there.

CONTENTS

FOREWORD

As a child, all I ever wanted to do was to take care of animals and visit zoos! So growing up in Knoxville made it easy on my parents (they didn't have to drive far to keep me happy!).

I remember Old Diamond, my favorite animal at the zoo . . . this huge elephant really captivated me. Still today, I can remember the look in his eye, his shuffling gait, the sounds he made—and yes—his odor! Therein lies the importance of zoos and animal parks—live animals have the ability to inspire people, both young and old, to care for them and hopefully to become conservators for the future!

It didn't take long before casual zoo visits didn't cure my intense "fix" for wildlife. Fortunately I was lucky to meet Dr. Warren Roberts—a local veterinarian that also happened to be the zoo's vet. At 11 years of age, I proudly told Dr. Roberts that I wanted to volunteer for him and do whatever he needed, just to be around the animals. Whatever made him accept this invitation from a wild-eyed kid I'll never know, but I'm glad he did! Obviously I became enamored with the work, and by the age of 16, my destiny was decided—I would enter the "Zoo World" as a career, starting as a zookeeper.

Years later, Dr. Bill Patterson and I founded the Appalachian Zoological Society when Guy Smith was the zoo director. Thanks to Guy, Bill, and the City of Knoxville, the zoo began its great turnaround. Today zoo director Jim Vlna leads the way as the zoo enjoys being the largest year-round attraction in Knoxville. Widely known for its large and diverse herpetology collection, it is involved in 30 of the Association of Zoos and Aquarium's Species Survival Programs. The Knoxville Zoo is proud to be a world leader of breeding red pandas and was the first zoo in the western hemisphere to successfully breed African elephants.

So yes—I definitely have many fond memories of the zoo and am happy to follow its continued success, especially since it fueled my desire to work with animals.

Many of us have second thoughts about our careers, but once I became involved with the zoo, I never looked back. I pursued my dream with fervor, as I still do to this day . . . and it all started at the Knoxville Zoo!

Jack Hanna

—Jack Hanna
Director Emeritus, Columbus Zoo,
and television's popular Wildlife Correspondent

Introduction

Perhaps someone's pet grew too large to keep at home, but however he came, in 1948, an American alligator named Al took up residence at the Birthday Park Zoo* at Chilhowee Park, which was run by the City of Knoxville. Much as it is today, Chilhowee Park was then a central location for community events, picnics, parades, and other activities. The park already had a wide variety of native animals—including chipmunks, squirrels, rabbits, and deer—so perhaps it seemed logical that there should also be a zoo on the premises. Or perhaps it was just convenient for people to bring their unwanted domestic or exotic creatures to the park and leave them with the willing yet untrained personnel to care for them. Whatever the reason, the alligator was left at the park, and many other animals followed, including ducks, lions, groundhogs, pigs, and monkeys.

While most of the early animals were donated, an animal was occasionally purchased as well. Sometimes they weren't purchased for viewing. A mule named Daisy was purchased around 1953 for the meat she could provide the carnivores. After Daisy proved she was capable of working for a living, though, she was put to work hauling a small cart of items around the park and up steep hills when necessary, and at other times, she served as an attraction in the children's petting area. When she finally died of natural causes several years later, she was then fed to the carnivores, serving her original purpose in the end.

Daisy's story is pertinent because it is important to note that it has always been difficult to sustain the zoo. Enclosures, or "habitats" as they are now called, must be built and maintained. Money has to be raised for expenses such as electricity, gas, water, repairs, and salaries. And of course, animals must receive vaccinations, medical check-ups and treatment, and stimulation. They must also be fed—frequently—which costs money. All the expenses for the zoo must be paid for in some way. An animal in captivity cannot care for itself, and it relies on those around it for shelter and food. Sometimes a captive animal even relies on its keepers for simpler tasks such as grooming.

No precise records with early zoo statistics have been found, but we do know that by 1970, the city was planning to shut down the small zoo and offered the animals for purchase. The sale of one little lion cub changed the history of a family, the zoo, and Knoxville forever.

Guy Smith, a local television executive, saw an article in the local paper that stated the city was selling two little lion cubs. Having wanted a lion club since he was a small boy, he came up with the $150 and took his companion home. Though he made life quite interesting, all went well until the lion cub, named Joshua, approached his first birthday. Smith and his wife, Patty, knew that they couldn't keep Joshua in their residential home much longer, but they also didn't want to turn him over to the zoo in its deplorable condition.

Through a series of events detailed in his book *A House for Joshua: The Building of the Knoxville Zoo*, Smith explains how he and his family did take Joshua to the zoo and leave him there in a cage. They visited so often that an employee in the mayor's office asked Smith if he would consider running the zoo. Fred Little, the zoo's head keeper at the time, and Smith seemed to get along well, and there was much work that needed to be done at the zoo. Smith wasn't interested in city politics or making friends just for the sake of making friends. He had a growing interest in the welfare of all the animals at the zoo, and he was definitely interested in the needs of his lion.

Most people acquainted with the Knoxville Zoological Gardens will say that this is when the zoo truly began. Smith took a few animals and cages at Chilhowee Park and over several years created beautiful zoological gardens. He convinced the city, local citizens, and even the humane society that this could be a great zoo if everyone believed in the project, put their support behind it, and helped it grow. This is exactly what happened.

When Smith came to the zoo in 1971, the city and the humane society were literally ready to close the zoo's doors. Just months before Smith came, several monkeys had been ordered to be put to sleep because one had contracted tuberculosis. Other animals had died. Old Diamond, an elephant who had been donated by Ringling Brothers Circus in 1963, had no adequate living space and tore down the wall to his elephant compound regularly. Soon after Smith became the zoo's director, at a pay rate of $1 per year, he began talking with local newspaper and television executives and people on the city council to gain support for the zoo. He joined the AAZPA (American Association of Zoological Parks and Aquariums), began bringing in animals that would attract the public, and focused on improving the image and the reality of the zoo. Smith saw early on how important it was to preserve endangered species, and the Knoxville Zoo became—and has continued to be—a premier zoo for its work in this area.

The history of what is now the Knoxville Zoological Gardens began with Guy Smith in 1971. This book will record those very important early years of Smith's leadership, as well as the changes and events that have taken place since then. The history of various animals, educational programs, activities, special events, and leadership will be covered.

The Knoxville Zoo has come a long way and will continue to change, grow, and thrive. As long as there are animals, particularly threatened or endangered animals, that need to be protected and as long as there are people interested in the beauty of nature and the creatures that live in it, the zoo will persist.

*The Knoxville Zoo has been known by several names over the years, including Birthday Park, the Birthday Park Zoo, the Municipal Zoo, Knoxville Zoological Park, and Knoxville Zoological Gardens. In specific locations where applicable, I have used the name for that time period; otherwise "Knoxville Zoo" is used throughout.

One

The Birth of Knoxville Zoological Gardens

An American alligator named Al came to live at Chilhowee Park in 1948. There seems to be no record of how he came to be at the park, but it's likely that he was someone's pet and that he grew too large to care for at home. Donating pets is not unacceptable, but taking in donated exotics can perpetuate an unacceptable trade. Holding certain species in private is illegal; donating, however, is not.

In 1970, Guy Smith made a decision that would change not only his life, but the lives of countless children, adults, and animals. When he found out that the City of Knoxville was planning to sell the animals from the local zoo, he promptly went to purchase this adorable lion cub whom Smith named Joshua. (Photograph by Michael Fouraker.)

After Guy Smith's lion, Joshua, grew too large to continue living in a domestic household, he was taken back to the zoo. Smith visited Joshua every day until finally a city employee asked Smith if he might be interested in taking over the zoo. Smith did become acting director at $1 per year, while holding his full-time job for three years.

Patty Smith not only helped her husband, Guy, take care of Joshua, but when he came home and told her that he wanted to run the whole zoo, it appears that she fully supported this decision as well. Over the years, she cared for many young animals in her own household and helped the zoo grow into what it is today.

Smith knew that for the zoo to survive, it must have animals and exhibits that attract visitors. Primates are one of the most popular exhibits at zoos, so Smith began making plans to have the monkey house revitalized. After a local lumber company donated materials and Civil Air Patrol cadets donated labor to complete the monkey house, Sulawesi macaques, like this one, were brought to the zoo in 1971.

Lincoln Hurd (left) and Bob Mize welcome two new members to the zoo's growing animal collection in May 1971. Like many other new arrivals, these jaguar cubs, Inca and Aztec, went to live at Guy and Patty Smith's house for a short period of time before being introduced to their new home at the zoo. (Photograph by Mickey Creager.)

Jaguars are the largest cats in the western hemisphere, and unlike other big cats, except tigers, they enjoy being in water. Here Robert Ambuser carries Inca and Aztec, the two wet jaguar cubs, back to their enclosure in July 1971. The jaguar exhibit became the snow-leopard enclosure in August 1989. (Photograph by Hugh Lunsford/ *Knoxville Journal.*)

Old Diamond (also known as Diamond, Ol' Diamond, and Ole Diamond) was donated to the Municipal Zoo at Chilhowee Park on April 1, 1963, from the Ringling Brothers Circus. With no accommodations prepared, the African elephant was originally chained in the ball field until workers could construct a small concrete building and an outside yard. Bull elephants have a reputation for being difficult to handle, and Old Diamond was no exception (this is part of the reason the circus left him in Knoxville). Old Diamond proceeded to tear apart his enclosure at the zoo every chance he could get. It is estimated that Old Diamond had been born in 1946, making him about 26 years old when this photograph was taken in 1972. He lived a productive life—performing as a circus animal, siring the first two African elephants born in the western hemisphere, and delighting thousands of visitors at the zoo—until he died at the Knoxville Zoo on September 10, 1980.

Creating exhibits that give visitors a more realistic view of how animals would coexist in the wild is common among zoos today. The Knoxville Zoo was already exhibiting animals in communities, but, by today's American Zoo and Aquarium Association's Standards, an Asian elephant would not share an enclosure with African animals, as pictured here in 1975.

Guy Smith came up with some very creative ways to obtain money to sponsor new animals for the zoo. When the zoo locked local WKGN disc jockey Eddie Beacon in a cage, fans were asked to make donations for his release. There was much fanfare about Beacon's "lockup," and in the end, the radio station raised $1,400 to buy the zoo's first tiger.

Once money had been raised to purchase the zoo's first Bengal tiger, Guy and Patty Smith traveled to the Jackson Zoological Park in Mississippi to pick up an available cub. Tigers, like most animals and people, can remember caregivers fondly for years to come. Sindhu, pictured here with Guy in the early 1980s, snuggles up to the fence in a display of affection. (Photograph by C. Suddarth Kelly.)

Many Bengal tiger cubs followed Sindhu. Pictured here are Skip Hollingsworth (left) and Lynn Cole with Asthuna Sundari. The young tiger cub arrived on December 6, 1973, as a purchase from the Washington Park Zoo in Indiana. Sundari died suddenly on February 14, 1980.

Mike and Carol Fouraker began working for the Knoxville Zoological Park in 1973 and 1974, respectively. They worked in many different areas and eventually both became curators. When Ted Beattie left, Mike served as acting director for a short period of time. The couple left the zoo in 1993, when Mike went to serve as director of animal collections of the Fort Worth Zoo in Texas.

Zookeeper Pat Parks pauses to visit Marvin the Rhinoceros. Protected animal contact has evolved over the course of zoo history. Zoo keepers have always used a combination of hands-off and hands-on management to work with animals, and this continues today, as zoos use touch as a form of positive reinforcement in formal behavior-management programs.

As part of his plan to improve the zoo, Guy Smith began looking for ways to obtain new animal species. He decided to join the American Association of Zoological Parks and Aquariums (AAZPA), which has since been changed to the Association of Zoos and Aquariums (AZA), so that he could receive member benefits, including a monthly newsletter and an animal-surplus list. Through this list, Smith could see which zoos had a surplus of particular species, and he could request those animals. The Knoxville Zoo first obtained red pandas in 1977, and in 1998, the AZA awarded the Knoxville Zoo the coveted Bean Award for its work in breeding and conserving red pandas. Sylvia, pictured here, was born June 14, 1992, and the Madrid Zoo donated her to the Knoxville Zoo in June 1995. She had 11 cubs born at the Knoxville Zoo and died here June 1, 2005. (Photograph by Betty Wasserman.)

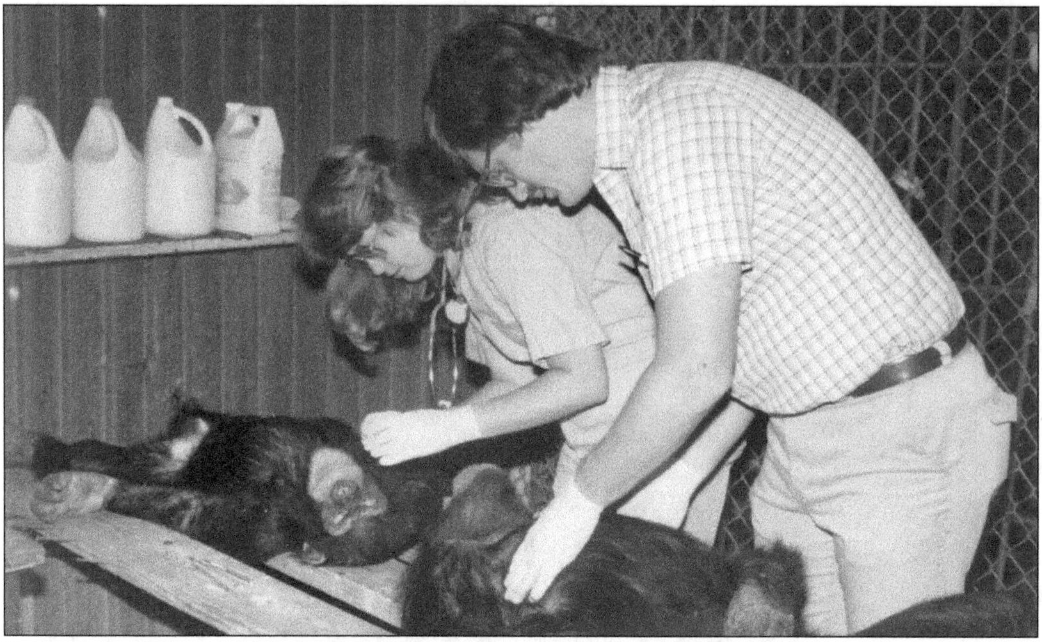

In 1976, Smith made mutually beneficial arrangements with the newly formed University of Tennessee College of Veterinary Medicine to provide animal medical care in exchange for hands-on experience. This photograph, taken in the early 1980s, shows Drs. Andrea Lenhardt and Jim Jensen performing routine physical examinations on two of the zoo's chimpanzees.

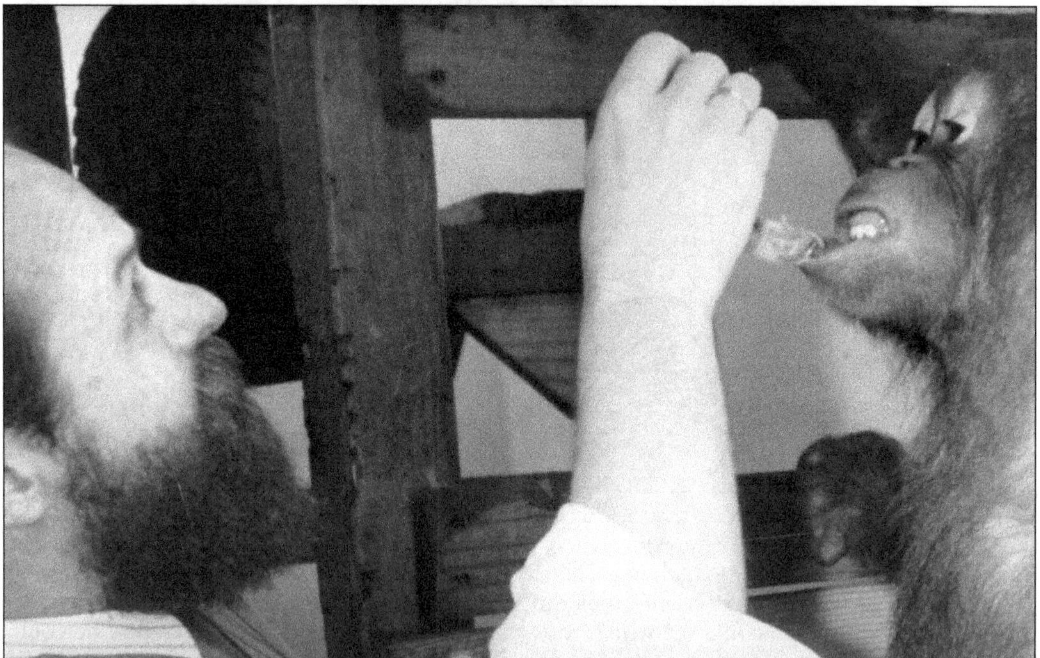

Dr. Ed Ramsay, one of the consulting veterinarians at the zoo, feeds an orangutan. The veterinarians who serve the zoo from the University of Tennessee perform all types of health-care services on all species of animals.

About 50 percent of the veterinarian services performed on zoo animals are preventative; 20 percent are service-type activities (chimpanzee and other animal introductions, for example); and of course there is treatment if the animals are ill or injured. Picture here are students from the University of Tennessee College of Veterinary Medicine examining a gorilla in March 1993.

The zoo is constantly making changes and expansions for the animals and improvements for its guests. In this image from *c.* 1974, the hippopotamus barn is to the right of the photograph and the giraffe barn is in the background just left of the center. This area is now the elephant yard.

A large collection of hoofstock was added to the zoo's animals in 1974. This is significant because it greatly increased the numbers of animals, which typically increases visitors. The hoofstock pictured here are eland.

Many of the animals included in the large hoofstock collection that was added in 1974, such as the white rhinoceroses pictured here, also became pioneer animals in the zoo's threatened or endangered breeding program. In this photograph from August 1974, the rhinoceroses are becoming acquainted with their new surroundings. (Photograph by Terry Moore, *Knoxville News-Sentinel.*)

Following the addition of hoofstock in 1974, the zoo added new walkways for visitors. In this image, taken *c.* 1975, you can see visitors at the right trying to get a glimpse of the new plains area, where giraffes, zebras, and other hoofstock animals were housed. The barriers around the trees are in place to keep the animals from eating them. The building in the background to the left is the old hippopotamus barn.

The old zoo entrance was located between the parking lot and the current Kids Cove. While that entrance area is occasionally used for activities such as Boo at the Zoo!, the actual building was torn down in 2000. About 600 dump-truck loads of dirt were then brought in to raise the area for Kids Cove to make it accessible for all visitors and to fill in the grotto moats.

When Diamond first saw his intended mate, Toto, he pushed her around a little bit before he ultimately accepted her. African elephants are an endangered species, and it was hoped in the late 1970s that breeding would occur in captivity in order to propagate the species. The pair finally did breed and in 1978 became parents to Little Diamond, the first African elephant bred and born in captivity in the western hemisphere. This was an amazing achievement for the elephants and the zoo, ensuring a place in zoological history for both. In this adorable picture, Little Diamond is pictured with her mother, Toto, plodding across the elephant enclosure. The baby elephant grew quickly, as all babies do, and she now resides at the North Carolina Zoo.

The first two African elephants born in captivity in the western hemisphere were born at the Knoxville Zoo. Shortly after the birth of Little Diamond (page 22), Hillary was born just two months later to Sapphire, who came to the zoo shortly after Toto, Little Diamond's mother. Diamond (page 13), the huge bull elephant, sired both. Both baby elephants are pictured here with Toto in 1978.

Petunia, an African elephant, poses for the camera—as she was trained to do—while giving some delighted children a ride. The zoo no longer gives elephant rides for safety reasons.

Education has always been a priority at the zoo. Pictured here from left to right are Frank Huskey, Guy Smith, and Jerry Reynolds with Little Diamond, an African elephant; Misty, a llama; and Prince, a donkey. (Photograph by Hugh Lunsford, *Knoxville Journal.*)

When the World's Fair was scheduled to come to Knoxville in 1982, zoo personnel hoped to see some of the crowds. Opening week attendance at the fair was 387,697 people. Instead of increased crowds, however, zoo visitor numbers dropped 47 percent in May (the first month of the fair) and continued to stay down until the fair was over in October. (Photograph by Brent Moore.)

After Guy Smith learned that he had cancer, he essentially hand-picked Ted Beattie to take over the zoo for him. The board approved this decision, and when Smith died in 1987, Beattie became the second zoo director. Ted Beattie and his wife, Penny, took over beautifully where Smith left off. The couple used their brilliant marketing and promotion skills to put the zoo on the map and make all the improvements and achievements known to the public. The zoo also added many new departments under Beattie's direction, including the formal Education Department, a Visitor Service Department, and a Grounds, Maintenance and Horticulture Department. Conservation efforts at the zoo were enhanced and recognized. Beattie continued to bring in animals, raise funds for the zoo, and raise Knoxville businesspeople's and leaders' awareness of the need to support the zoo. Beattie is now the executive director of Shedd Aquarium in Chicago.

Ted Beattie (right) pets some donkeys in the children's petting-zoo area while speaking with Mike Fouraker. (Photograph by Betty Wasserman.)

A rare white Bengal tiger cub wrestles her sister—a normally colored orange Bengal tiger cub. Both have black stripes. The seven-week-old cubs were only two of the 75 Bengal tiger cubs in captivity in the United States. The Columbus Zoo loaned the cubs to the Knoxville Zoo for a two-week visit, July 15–29, 1987.

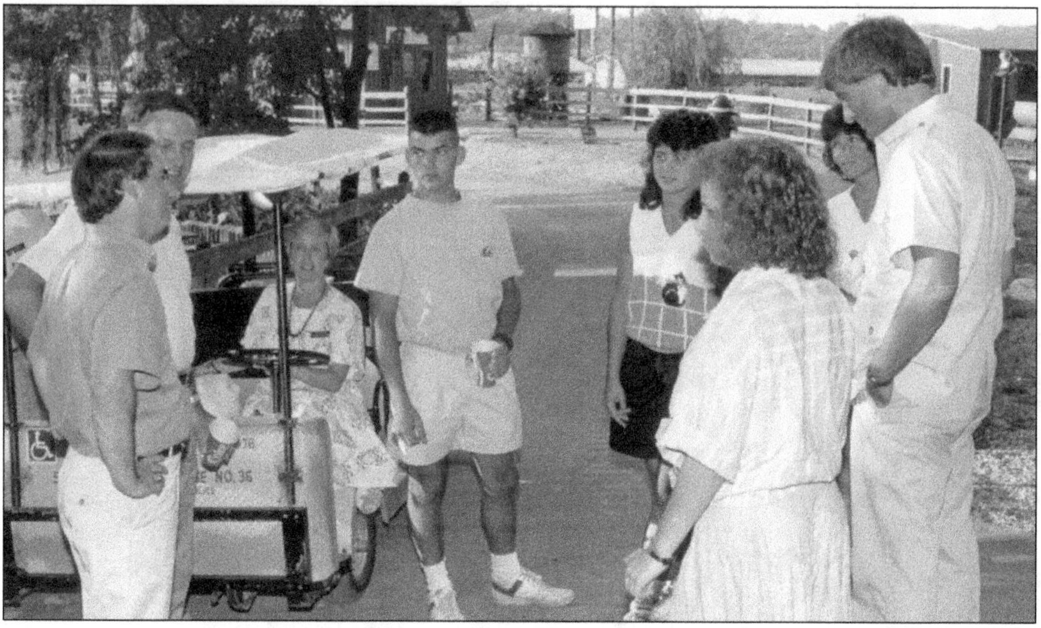

Jack Hanna has always loved animals, and at age 11, he began volunteering for veterinarian Dr. Warren Roberts, who allowed Hanna to help care for the animals at Knoxville's Municipal Zoo. Above, Hanna (far right) talks with Knoxville Zoo director Ted Beattie (far left), Penny Beattie (seated in the golf cart), and other zoo employees around 1987. Hanna has donated several animals to the zoo and frequently returns to visit.

When he was young, Jack Hanna's parents and local veterinarian Dr. Warren Roberts encouraged his love for animals. Into adulthood, Hanna knew he wanted to work with animals and became one of their leading advocates. Today Hanna continues to promote awareness of animals and their needs as director emeritus of the Columbus Zoo and as television's popular Wildlife Correspondent. (Photograph by Rick Prebeg/World Class Images.)

K'Bluey Koala came on loan from the San Diego Zoo for a month in June 1988. Keeper Jane Gilbert is shown here preparing for the return flight to San Diego at the end of the summer. Delta Airlines was one of the sponsors of the koala.

Though many people think of the koala as a "teddy bear," it's actually not a bear at all. It's a marsupial native to Australia and has a pouch in which to raise its young. Koalas live in the upper branches of eucalyptus trees, eating a diet solely of eucalyptus stems and leaves. These adorable little creatures have few natural predators other than humans, who slaughter the koala for its gray pelt. (Photograph by Tom Lane.)

Though the board of directors conducted an exhaustive national search, they ultimately selected Patrick Roddy, whom they knew as one of their own board members, as new zoo director when Ted Beattie went to Chicago's Shedd Aquarium. Supported by the board of directors, Roddy spearheaded a major capital campaign, raising funds to rebuild one-third of the zoo. The campaign brought in amazing results, including the building of Zoo Drive, a new entrance location named the East Tennessee Plaza to honor Patrick M. M. Roddy by the Pat Roddy III family, Black Bear Falls, Gorilla Valley, and Chimp Ridge. With a financial background, Roddy recognized the need for financial stability. In addition to the building of the major exhibits, Roddy also supported the construction of major amenities such as Safari Grill and the Zoo Shop. Roddy was director of the zoo from 1993 to 2000, during which time he encouraged zoo staff members to think of the zoo as the world-class institution it is. Roddy is shown here with one of the zoo's giraffes in 1994. (Photograph by Betty Wasserman.)

One of the things zoo directors Ted Beattie and Patrick Roddy both did was encourage awareness of the zoo's accomplishments. They wanted the community, including political leaders and business owners who could donate money, to know about all the wonderful things going on at the zoo. Some of the promotional activities, such as this Do Dah Parade from the early 1990s, were quite creative.

The East Tennessee Entrance Plaza, sponsored by the Roddy family, reflects the thoughtful planning found throughout the zoological gardens. On a small brick island near the entrance area, there is also a small glimpse of the many trees and plants that visitors will see in the zoo. The entrance area includes a gift shop, the ranger area, membership and ticket windows, and the admission gates. (Photograph by Jim Vlna.)

During his years as zoo director from 1993 to 2000, Patrick Roddy focused on building amenities for visitors to bring in money for zoo operational expenses. Pictured here is the Oasis refreshment stand, built in 1991. Currently, thanks to money from gate admissions, food concessions, memberships, gifts and souvenirs, and rides, the zoo is now 86 percent self-supporting. The remaining 14 percent of operational dollars comes from the city and county governments and private contributions.

In this construction photograph from the early 1990s, the *Knoxville News Sentinel* Amphitheater is shown. The popular Bird Show, which takes place here, presents free-flight natural behaviors and has educated and entertained millions over the years.

Jim Vlna began working at the Knoxville Zoo in 1991 as operations manager. One of his early responsibilities was to build the Master Plan developed under Ted Beattie's watch and rebuild a failing and mostly nonexistent infrastructure. Vlna was promoted to deputy director by Patrick Roddy, and when Roddy resigned in 2000, Vlna was unanimously chosen by the board of directors to replace him as zoo director. The decision was met with unanimous support at an all-employee meeting. Much like the zoo's first director, Guy Smith, Vlna loves the animals he cares for. He brought to the job a specialized knowledge of zoo operations, construction, guest services, business practices, budget preparation, and future planning. Vlna continues to strengthen the foundations that have already been laid, begin new construction programs that will improve the zoo, and solidify the zoo as an outstanding regional and community asset.

The Clayton Family Kids Cove is the only themed playground that also features animals of the Appalachian region. The exhibit, which opened on April 2, 2005, is approximately three acres designed around the theme of an early-1900s Appalachian farm. The Kids Cove has an animal-contact yard, a climbing playground, a giant sandbox, a water-play area, and the only carousel in Knoxville featuring custom-made animals such as a red panda and an alligator. There is a huge bird's nest to play in, a rock-climbing wall for kids, and also a fake cow to practice milking. The area is geared primarily toward children ages 1 to 10, but children of all ages would enjoy this spectacular Appalachian-themed playground at the zoo. There are many people and businesses that donate money to the zoo, and these gifts are greatly appreciated. But even with all these donations, major exhibits at the zoo wouldn't be possible without major sponsors. Kids Cove was sponsored by the Clayton family. (Photograph by Jim Vlna.)

Two

ENCLOSURES VS. HABITATS

This picture shows a view of the former bear dens behind the grottos (the front view can be seen on page 42). This location is currently the barn in the Clayton Family Kids Cove. Although the bears have since been moved to a wonderful new habitat, the dens shown here (now refurbished) are still used to hold guinea hogs and beavers.

It takes a lot of effort from many people to build new habitats in a zoo. Some of the activities that must take place are fund-raising to build the habitat, designing the area, constructing and testing the habitat, moving the animals, introducing the animals to a new living area without stressing them, and grounds keeping of the new area. Here is a photograph of Gorilla Valley being constructed in 1992.

Gorillas are very intelligent animals. Like all animals at the zoo, gorillas receive daily "enrichment," which is defined as providing activities and stimulation to encourage species-typical behaviors. Enrichment might consist of bananas hung from the top of the cage for the gorillas to get, tree limbs that the gorillas can tear apart to eat, or even a tub of water to explore. Mind-stimulating toys and activities are important to the well-being of all animals in captivity. (Photograph by Paul W. Faust.)

Frequently when new exhibits or improvements to exhibits are needed, the zoo has come up with creative fund-raising tactics. When the otters needed a new area to live, the zoo sold tiles to fund the construction of the otter exhibit. The Wall of the Wild opened in spring 1995. The promotion was so successful that people still ask about having tiles added.

River otters originally came to the Knoxville Zoo in 1981, and they lived in an exhibit close to the Ravi and Burma tigers. The otters pictured here were brought in specifically to live in the current otter exhibit with a very naturalistic setting. The zoo received five otters in May 1994. They were housed in the clinic until the exhibit opened the next year. (Photograph by Diane Fox.)

This beautiful new barn was built in 1992 as the children's petting-zoo area. When Kids Cove was constructed, the petting area moved there, and the llamas and alpacas remained in this barn, which was donated to the zoo by Barna Log Homes.

Kids Cove is one of the newest attractions at the Knoxville Zoo. While cold, dull animal enclosures have changed to beautiful habitats, the children's area has changed from a simple petting area to a remarkable playground surrounded by some of the area's most fascinating animals.

These round, wire cages, also known as "circle cages," were actually designed as corn cribs. At some point, someone decided to house animals in them, and the zoo purchased several of these in the early 1970s. The larger cages were used to house exotic cats, such as leopards or lynxes.

The smaller circle cages, shown from farther back in the top photograph and closer in this photograph, were used to house primates. Many animals have been moved to larger, more natural habitats.

Patty Smith visits a baby black leopard that she helped raise. The leopard is in one of the old large circle cages, originally constructed as corn cribs. Though Smith was comfortable with the leopard, and the leopard most assuredly remembered her fondly, even animals that are hand-raised are still wild animals. Exotic animals should always be treated with caution, even by those who are very comfortable with them.

Using many gallons of water and having frequent leaks, the marine-animal complex built in 1984 provided a less-than-adequate quality of life for the animals it housed and was torn down in 1998. The building to the left still serves as the penguin house, but the area to the right has been dismantled. It was located where Black Bear Falls is now. The Marine Animal Complex was featured in an article in *Southern Living* magazine in 1985.

Sea lions were one of the favorite attractions in the newly built Marine Animal Complex. A male named Alpha arrived in August 1984, and a female named Dolly came in 1985. The pair had a baby in June 1988. Then, in June 1989, they had Kassi. Dolly died in 1994, and Alpha was relocated to the Indianapolis Zoo in November 1997.

The zoo had 13 penguins and one on loan as of December 2006. Zoos loan animals through a nationally organized breeding program called Species Survival Plan (SSP). These programs, administered through AZA, promote responsible breeding to maintain genetic diversity in a species. (Photograph by Evelyn Johnson.)

This was one of the grottos built to house black bears, grizzly bears, and sun bears. In this picture, you can see visitors viewing baboons, which were moved to the grottos after the zoo stopped exhibiting sun bears and grizzly bears. Black bears were still there until a new habitat was built for them.

Pictured here is an Asian black bear sitting on the steps in one of the old grottos. The Asian black bear was declared a vulnerable species on the Red List of the International Union for the Conservation of Nature and Natural Resources (IUCN). The Knoxville Zoo had these beautiful bears from 1983 until 1987. (Photograph by Louise Cox.)

Black Bear Falls is one of the best bear exhibits in the nation, providing both horizontal and vertical space for the bears to roam. The spacious, open-air exhibit is home to the zoo's black bears. It has been designed to simulate the bears' mountain habitat with natural-looking trees, tumbling waterfalls, and flowing streams. The exhibit also offers guests an inside view of the animals' den. Black bears were moved from the old grotto area to Black Bear Falls on September 8, 2000.

Like many of the old enclosures, the chimps' old enclosure had lots of fencing, few or no trees, and a ceiling. On the next couple of pages, you will see what a wonderful change was made when the zoo went from an enclosure to a more natural habitat.

Chimps were moved from the old chimp enclosure to Chimp Ridge on March 31, 1993. The Great Ape building serves Gorilla Valley and Chimp Ridge as a night house for the animals. The City of Knoxville, under the leadership of Victor Ashe, contributed a large percentage of the money for this project. Ashe also contributed personally to the zoo. The Alcoa Foundation supplied money for gorillas and provided additional staff training.

Many of the rocks around the zoo, particularly those in exhibits, are actually human creations. They start out as wire formations, then they are covered with mesh, then a special grade of concrete is blown onto the mesh with a hose. The zoo's exhibit-design staff create all zoo exhibits in-house.

Once an exhibit has been created, it has to be thoroughly tested to make sure it is safe for the animals. It also has to be tested to ensure that it is escape proof. Here professional climbers test Chimp Ridge to see if they would be able to climb up and out of the exhibit if they were living inside and wanted to get out.

Pictured above is part of the newly constructed Gorilla Valley. Notice the difference between a habitat, pictured here, and an enclosure, such as the one at the top of page 44.

Though the old grasslands area that housed hoofstock did have some trees, there were not many. There also wasn't much grass. The animals were well cared for, but in addition to good health care, adequate shelter, and plenty of food, it was discovered that animals fared better in surroundings more like the ones where they would be found in the wild.

Scott Niswonger sponsored the Grasslands Africa! exhibit at the zoo. Now hoofstock can roam through an exhibit that is designed to emulate their native Africa. A favorite animal in the Grasslands Africa exhibit is always the giraffe. (Photograph by Jim Vlna.)

A Kori Bustard runs through its cage. The zoo has had only two Kori Bustards. Both arrived in 1987, and both are females.

The new bird sanctuary has less visible fencing, more trees and plants, a water feature, and plenty of room for the birds to fly.

Both of these images show the old elephant yards. The photograph below shows what later became the elephant-ride area. The building in the background of the bottom photograph is the old elephant barn, near where the carousel and the Kids Cove restroom now sit. This image was taken *c.* 1979.

Playing around in the pool was a favorite pastime for the elephants, particularly the little ones. Pictured here are three young elephants in the late 1970s taking turns flipping over in the mud.

The elephant house provided shelter for several elephants at one time. This is where Toto and Sapphire both gave birth to the first two African elephants born in captivity in the western hemisphere.

The Stokely African Elephant Preserve, which opened in 2002, was sponsored by the William B. Stokely Jr. Foundation. The exhibit is designed to give the elephants lots of room and allow visitors to see the impressive animals, yet natural barriers keep elephants and visitors safe. (Photograph by Jim Vlna.)

In this 1970s photograph, three lion cubs peer out their fenced-in enclosure. Before Smith took over, a 1955 newspaper article stated that the Municipal Zoo traded two lion cubs for three monkeys and a South American porcupine. Today animals are exchanged through the SSP to ensure genetic diversity, to accommodate a zoo's space constraints, or to occupy new exhibit space.

Sonja was an Asian lion purchased from the Tierpark Berlin Zoo on December 30, 1977. She was born in May 1976. Sonja lived at the Knoxville Zoo until December 14, 1988. Sonja gave birth to seven cubs while she was here, and all but two survived. As evidenced in this photograph, it appears that Sonja was a diligent mother.

A young male African lion rests on a rock in the new Pridelands exhibit. African lion males are considered mature at about two years of age.

Lions live in prides, or social groups, made up of a dominant male lion, several lionesses, and their offspring. When a male cub reaches adulthood, he is expected to go off and find his own pride. A male lion and two lionesses sit in the shade of a tree in this early photograph from the Knoxville Zoological Park.

Three

Animals

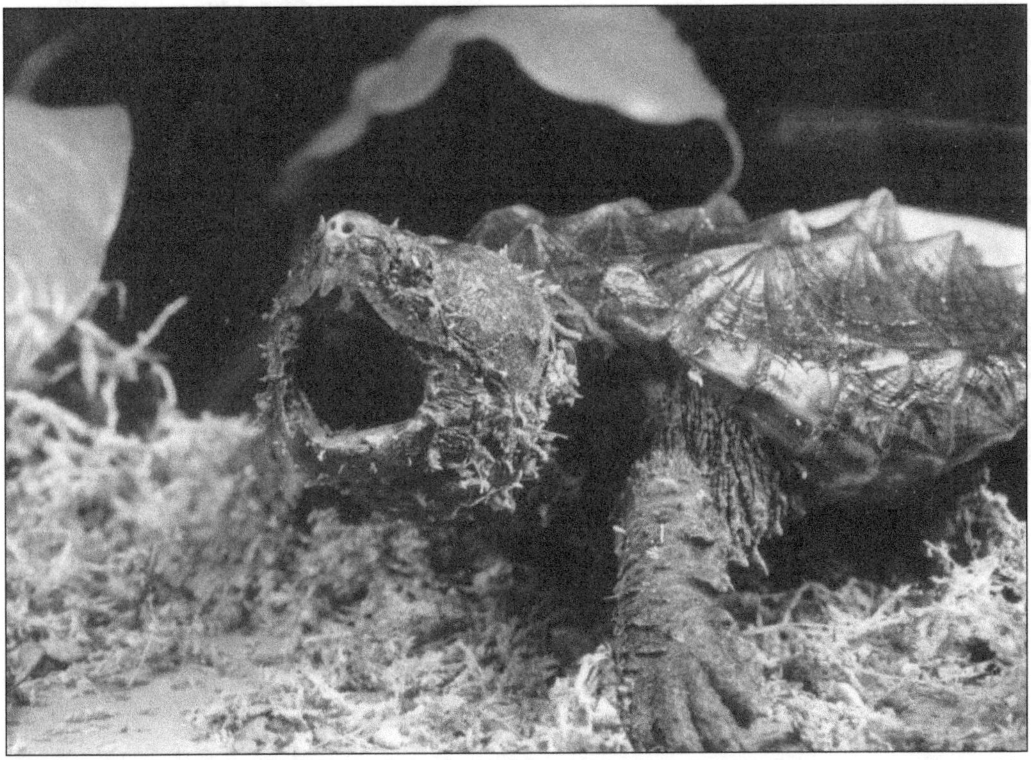

The alligator snapping turtle is the largest freshwater turtle in the United States and possibly the largest in the world. The International Union for Conservation of Nature and Natural Resources (IUCN) has this turtle listed as a threatened species, which means that they could become an endangered species in the near future. Once completed, the zoo's new reptile facility will continue its focus on breeding threatened or endangered reptile species and on conservation of reptiles and amphibians. The Knoxville Zoo houses the ninth largest reptile collection in the United States.

The Arabian oryx is a good example of the success of captive breeding programs through AZA's SSP program. The oryx had been hunted to extinction in the wild by 1972, so focus was placed on breeding the animals in captivity. Over 100 oryx have been released back into the wild and populations have reproduced, but the animal is still endangered, mostly because of illegal captures and sales. There is speculation that these animals are the original "unicorns" because, if you look at them from the side, you see their two horns as a single straight, pointed horn (as is the case with the second oryx from the left). (Photograph by Betty Wasserman.)

Binturongs, also known as bearcats, are really neither bears nor cats. They are relatively small animals—about the size of a medium dog—and dwell in the forest canopy in the wild. Pictured here is keeper Karen Webster giving the zoo's binturongs a snack. What you cannot see in the photograph are the binturongs' prehensile tails, which can be used like a fifth hand.

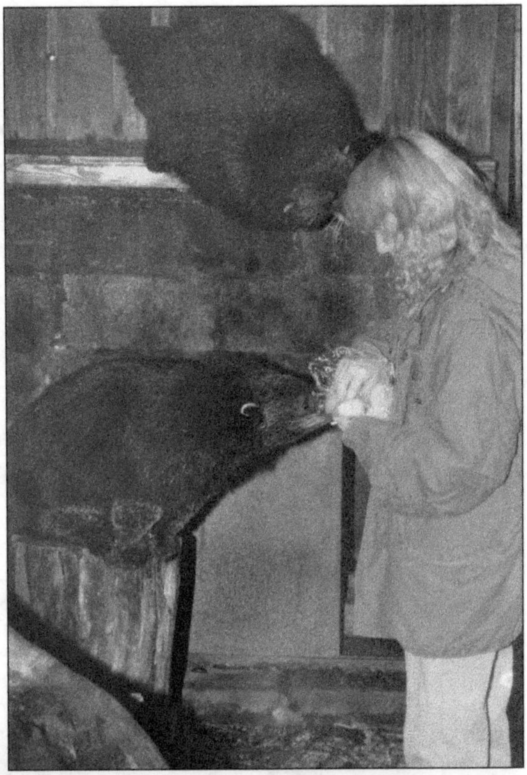

Roseate spoonbills are named for their bright pink plumage and their long, spoon-shaped bills. Their plumage is quite magnificent, and visitors definitely should pay a visit to these bright birds while touring the zoo. The only spoonbill species to live in the western hemisphere, these birds can be found naturally along the Gulf of Mexico coastline and in southern Florida. (Photograph by Micah Haskins.)

This beautiful feathered animal is the national bird of the United States and is the only eagle with a range restricted to North America. The bald eagle, so named because of the head of white feathers atop its black body, is a majestic creature that builds huge nests in the upper branches of large trees. In 1971, an unknown assailant used a shotgun to kill one of the two precious eagles at the zoo. Not only was it a cruel crime, it was a federal offense because eagles are actually property of the federal government. An endangered species in the early 20th century, the bald eagle has made a comeback thanks to preservation and protection laws and is close to being removed from the endangered species list. The bald eagle exhibit opened in 1991.

Keeper John Croxton pauses to watch a bush dog, named Cravo, have a snack. Cravo came to the zoo with her partner, named Pancho, *c.* 1987, and the pair had two offspring that did not survive. Bush dogs are listed as a vulnerable species, mostly due to habitat loss. (Photograph by Betty Wasserman.)

Black Bear Falls, a beautiful exhibit that allows visitors to watch American black bears in a naturalistic setting, opened in 2000 and is sponsored by the Lucille S. Thompson Family Foundation. With the support of donors, the zoo is able to create and maintain superb exhibits that enrich animals' lives. Pictured here are Otis and Alvin at home in Black Bear Falls. (Photograph by Paul W. Faust.)

Bactrian camels pose with their babies in June 1979 (top photograph) and 1980 (bottom photograph). These unique animals have two humps, unlike one-hump Dromedary camels, which are more common. Camels remain one of the zoo's favorite animals, especially among young people. Camel rides are available at the zoo. (Top photograph by Dave Carrier, *Knoxville News Sentinel*; bottom photograph by Michael Patrick.)

Cheetahs are no longer found in many parts of their native habitats of North Africa and Southwest Asia due to encroachment and poaching. Unfortunately these fast cats are rapidly declining and are critically endangered in some areas. Cheetahs are the fastest land animals and can travel up to 70 miles per hour for short distances, enough to catch their almost equally fast-running prey.

A Chinese leopard named Chu Yuan Chung, pictured here, was sent to the zoo on loan in November 1973. The zoo obtained a pair of Persian leopards as permanent residents in 1979. These animals are now known as Iranian leopards. The pair, named Morris and Courage, was transferred to the San Diego Zoo in 1983.

In this photograph, Vinnie and Peter Capybara take their offspring out for a stroll. While most zoos assign names to all the animals, it is also common to add the common name as a last name for the animals, thus "Vinnie and Peter Capybara." Capybaras are the largest rodents in the world, and they can weigh over 100 pounds. Capybaras are excellent swimmers and are social animals that should be kept in groups. A social group normally consists of about 20 animals, including a dominant male, several adult females, and offspring. There are also subordinate males in the social group. Capybaras only live to be about 10 years old.

Chimpanzees can grow to be between two and four feet tall and weigh between 57 and 155 pounds. Of course females are smaller than males. While chimps can live to be 60 years old in captivity, they rarely live past 40 years of age in the wild. However, Cheeta, the star of the movie *Tarzan*, was still alive in 2006 at age 74. This is female "Moms" with one of her several offspring.

The Knoxville Zoo has had three types of cranes: East African crowned cranes, Florida sandhill cranes, and Sarus cranes (from 1988 to 1996). This picture shows a beautiful East African crowned crane, a species the zoo still had on exhibit in 2006.

In 1973, Shoney's restaurant donated $5,000 to the Appalachian Zoological Society. A portion of the money was used to purchase this four-month-old cougar to add to the zoo's animal collection. (Photograph by Hugh Lunsford, *Knoxville Journal.*)

This photograph was taken of the cougar at age eight in April 1981. Cougars are also known as pumas or mountain lions. They are solitary animals and can be found in various locations of North America. The animals can actually grow to be very large, up to 200 pounds and eight feet in length from nose to tail as adults. (Photograph by Keith G. Dayhuff.)

The drill, which is a type of baboon, is one of Africa's most endangered primates. They are Old World monkeys and are closely related to mandrills, but they do not have very colorful faces like mandrills do. Drills are mammals, meaning—among other defining factors—that they have fur or hair, specialized teeth, warm-blooded bodies, and mammary glands that enable them to nurse their young.

While some animals must be anesthetized to move them from location to location and others require cages with thick, steel bars, the tortoises just walk casually, and slowly, to their new habitat. The opening of Tortoise Territory in 1994 brought many onlookers, including guests, zoo staff, and media.

Einstein is a Congo African grey parrot that was hatched in California in 1987. Einstein was purchased by Steve Martin, an early Bird Show consultant, to be part of the show's collection. He later sold the bird to the Knoxville Zoo. Since then, she has been the star of the zoo's Bird Show and a popular "spokesbird" for the zoo. African grey parrots are one of the smartest bird species and are also known as one of the best talking/mimicking species. Einstein is extremely intelligent, even for an African grey. She has an extensive verbal vocabulary and is able to mimic animals and other sounds. It is believed her vocabulary exceeds 200 words and sounds. Of that, about 80 are on cue. Einstein has made numerous appearances on national television. (Photographs by Dick Dickerson.)

Two large African elephants stand with their keepers in the old elephant area, which was located in what is currently the Kids Cove. The Night Club is part of Kids Cove and features native nocturnal animals such as bats, raccoons, and skunks. Special lighting effects reverse night and day for the animals. This allows visitors to see them during their most active time of day.

Gelado baboons belong to a grass-grazing group of primates that were once widespread, but now only the Gelado baboons remain. Unlike other female baboons, whose genital area is red and swells to indicate estrous, the female Gelado baboon has a swelling on the chest that transmits information to males.

Gibbons originally came to the zoo in 1974. They were loaned to Opryland in 1990 but then returned to the zoo when Opryland closed in 1997. In 1999, Euel and Nipper Gibbon had a baby named Archie. Gibbons' average life span is only about 30 years, so it is pretty amazing that they had a baby in what would have been considered old age. Nipper is currently estimated to be 39 years old.

When he became director, Guy Smith began bringing in animals that visitors like to see at zoos, including primates and bears. The grizzly bear, first brought to the zoo in 1971, is a stunning, majestic animal. Colors range from yellowish to dark brown and there is usually a tip of white on the ends of the hairs. This white tip results in a frosted look, which gives the bear the name "grizzly." Grizzly bears can grow to be six to eight feet tall. They are opportunistic omnivores, eating whatever they can find, including berries and other fruit, vegetative matter, leaves, fish, other mammals, and insects, as well as bark, mushrooms, and roots. The bears will also eat garbage, which is not only unhealthy for them but also harms them by bringing them in close proximity to humans, which ultimately leads to their demise. Animals in zoos are on very strict diets and are fed only what is most beneficial for them, so visitors should never throw food or other matter into cages or habitats. (Photograph by Hugh Lunsford, *Knoxville Journal*.)

Giraffes are the tallest of all the land-living animals and can grow to be 15 to 18 feet tall. They can extend their height even more by using their prehensile tongues to reach leaves even farther away. A giraffe's gestation (pregnancy) lasts a little over a year (14–15 months), and when the calf is born, it actually falls the five feet or so to the ground, bursting its amniotic sac when it hits the ground. Amazingly, the little calves survive this drop and within a couple of hours are usually walking around as if they had always done so.

While female giraffes hang out in social groups with other females and young giraffes, or calves, males prefer social groups of other bachelor giraffes. Older male giraffes frequently lead solitary lives and only come back to the social group when it is time to mate. Giraffes are polygamous, meaning that the males usually mate with all the females in the herd who are fertile. Their size and distinctive coloring make the giraffe one of the most noticeable of Africa's animals, and most children can recognize these animals in the zoo at a very young age.

The hippopotamus got its name from a Greek word meaning "river horse." Perhaps this is because this huge animal is usually more comfortable in the water than on land. Hippos live in groups called "pods" or "herds" and can live to be 40 to 50 years old. Knoxville Zoo's two hippos, Jabba and Funanni, currently live in San Diego Zoo and have produced several offspring. (Photograph by Hugh Lunsford, *Knoxville Journal.*)

The Knoxville Zoo had brown lemurs from 1983 to 1991 and ring-tailed lemurs from 1985 to 1994. The lemur in this photograph is obviously a ring-tailed lemur. Lemurs have opposable digits (our pinky finger and thumb are opposable digits) on their hands and feet, allowing them to grasp tree branches and small objects.

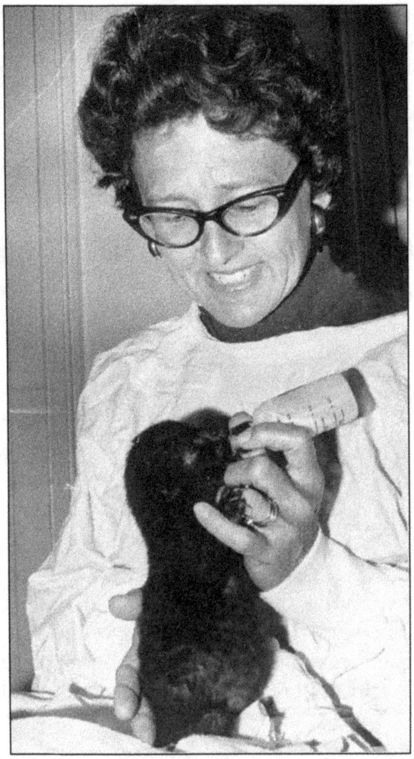

Shoney's restaurant donated $5,000 for two black leopards and a cougar in 1973. The black leopards were Cleopatra and Mark Anthony. The pair mated and had their first cub in May 1975. Unfortunately Cleopatra had difficulty feeding and caring for Aurelius, the cub, and she was near death when zoo staff decided to remove her from her mother. Pictured here is Patty Smith feeding the newborn (right). While she was staying in the Smith home, Aurelius was kept in an incubator. She lived at Knoxville Zoo for 17 years.

Knoxville Zoo has been recognized as having one of the finest reptile collections in the nation. They are currently working on an Amphibian Reptile Conservation Center to further establish their international reputation for the successful breeding of rare and endangered species of reptiles. Pictured here is a leopard gecko, which is just a scaly reptile.

The lynx is a gorgeous cat with a stubby tail and a tuft of hair on the tips of its ears. Their large paws are padded for walking on snow since their natural habitats are high-altitude forests primarily in North America. The lynx became a threatened species in 2000 and is thus protected by law from hunters and poachers. The wild cat is currently being reintroduced to the United States by the Colorado Division of Wildlife.

Pat Parks, senior keeper of hoofstock, visits two of his charges in January 1982. (Photograph by Hugh Lunsford, *Knoxville Journal.*)

Llamas are approximately four feet tall and are domesticated animals that come to us from Peru and Bolivia. The animals have been used in the mountainous regions there for thousands of years to carry goods over the rugged terrain. They are not ridden, however, and when they are overloaded or become angry at their drivers, llamas can be downright nasty, frequently spitting at the person who has upset them.

These adorable little creatures are called meerkats. They are mammals who, in the wild, inhabit all parts of Africa's Kalahari Desert. Meerkats have very organized social structures and travel in groups called "gangs" or sometimes called "mobs." They are quite small, growing only to about 12 inches in height, and they subsist primarily on insects, although they will also eat lizards, eggs, spiders, and other food as the opportunity arises. Though not immune to venom, meerkats are very quick and almost always succeed in seizing a snake behind the head. (Photograph by Micah Haskins.)

Miniature horses can be found in all parts of the world and are always less than 34 inches in height. They are domestic animals and normally live about 30 years, though some have been known to live longer. Miniature horses were not bred for riding or other practical uses but were originally bred as pets for European nobility in the 1600s. Pictured here is miniature horse April Sunshine.

The ostrich is the largest living species of bird and is one of the few flightless birds. Ostriches live in groups and travel with other grazing animals. While they can go without water for long periods of time, ostriches must swallow pebbles to help them grind up their food, mostly seeds, plants, and occasionally insects. Ostriches can live to be 75 years old.

Owls are solitary, nocturnal birds of prey. They hunt small mammals and other birds, and some also feed on insects and fish. Owls are found around the globe except in the coldest climates such as Antarctica and the most remote islands, where it is likely that they could never have migrated. Owls are known for their huge eyes, which allow them to see far away. Their night vision is particularly good. This is a snowy owl.

Pacco was a green-cheeked Amazon parrot. He was donated to the zoo by a private collector. It is not possible to see the beautiful colors of Pacco's feathers in this photograph, but he had vibrant green feathers covering his body with a bright red patch and a light blue patch on his head.

Though this stunning animal is frequently referred to as the peacock, that is actually only the male of the species. The correct name of the species is peafowl. The females are called peahens. The zoo had peafowl in 1980 and then again from 1985 to 2000. The birds are members of the pheasant family and are most famous for the males' bright, colorful tail feathers.

This Asian potbellied pig was received from the Columbus Zoo in exchange for a Sicilian donkey. Frequently zoos exchange animals when one zoo needs a particular species and another zoo is willing to trade. In addition, moves are sometimes necessary to fulfill the SSP goal of preserving a particular threatened or endangered species.

Knoxville Zoological Park received three polar bears donated from the Cleveland Zoological Park in Ohio in October 1973 with plans to expand the zoo. A polar-bear exhibit was opened in 1974. Guy Smith had a vision to create a better zoo—for animals and visitors. A Marine Aquatic Center was part of this vision.

This photograph was taken as part of the Zoo View photograph contests that were offered in the 1980s. Visitors were allowed to take photographs and submit their best for consideration in the contest. This particular image was taken by Bridget McAdams of Strickney, Illinois, in 1987.

Birds, such as the Peregrine Falcon pictured here, have been at the zoo since before Guy Smith's arrival in 1971.

The scarlet macaw can be found in the wild anywhere from South America up into Mexico. These particular birds can make good pets to someone who can commit to a bird that can live to be 70, has the money to purchase it and the huge cage that must come with it, and has the ability to provide food (mostly large seeds) and lots of time outside the cage for the bird. These birds can be dangerous to young children and require lots of attention, but scarlet macaws can be affectionate and can learn to talk well.

Red-tailed hawks are frequently used for falconry (which involves training hawks to hunt) in the United States. This practice does not have an adverse effect on the population, however, because the survival rate for young red-tailed hawks in the wild is actually much lower than it is for those in captivity. Also state and federal regulation dictates that the hawks must be released upon reaching maturity. Then these mature hawks actually have the opportunity to breed. This is a red-tailed hawk from the Knoxville Zoo in April 1983.

Sable antelope are one of the largest members of the antelope family. They are from Southern Africa. A sable's horns can grow up to 48 inches in length and can be very dangerous if they are charging. Sable antelope inhabit woodlands and grasslands and typically graze facing outward to keep an eye out for predators, such as lions.

This beautiful cat, called a serval, is actually quite common among the savannahs of Africa. The cat has been hunted to extinction in South Africa, however, by people wanting to exploit the animal for its attractive fur. Some attempts have been made to breed a domestic cat with the fur patterns of a serval, but this typically results in cats that are neither suited to wild living nor domestic life.

This adorable little cat is a snow leopard named Arant, who was born at the Knoxville Zoo on April 28, 1988. This picture of him was taken in July 1988. Arant lived at the zoo until March 5, 1991, when he was loaned to the Pittsburgh Zoo. He was later donated to that zoo.

The snow leopard is native to the mountainous areas of Central Asia. They are unique among the cats thanks to their white, almost silver-looking coat, and their long, bushy tail, which can be used to aid the cat in balance or to provide a warm cover for the cat's mouth and nose in very cold weather. (Photograph by Ron Warick Studios.)

Knoxville Zoo's Aldabra tortoises reside in a large spacious outdoor enclosure during the warm months of the year and have become visitor favorites. (Photograph by Bern Tryon.)

Adult female bog turtles are smaller than males and average about three and a half inches in shell length. The bright yellow patches on the head can be used to readily identify a bog turtle. (Photograph by Bern Tryon.)

Pictured here are the first three Madagascar Flat-tail tortoises hatched at Knoxville Zoological Gardens. This is a critically endangered species endemic to western Madagascar. This species has been in the collection since 1975. In May of 2002, Knoxville Zoological Gardens became the first zoo in the United States to successfully hatch this species. (Photograph by Michael Ogle.)

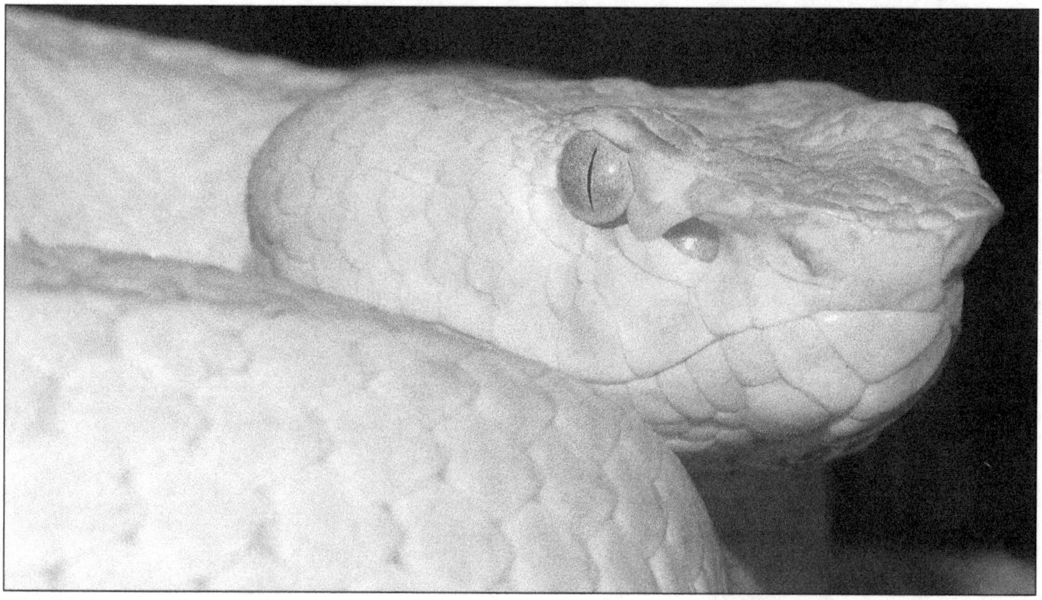

The Philippine pit-viper occurs in several color phases, primarily yellow or white. Shown here is a solid white phase. Little was known about the Philippine pit-viper when the Knoxville Zoo received a group of one male and three females in 1993. Since then, the zoo has bred them four times and is one of only three U.S. zoos to have been successful with this AZA Studbook species.

The Knoxville Zoo had Malayan sun bears from 1974 until 1990. Frieda was the first sun bear at the zoo, and she was soon joined by a female named Theta and a male named Yogi. A sun bear is pictured here in one of the old grotto enclosures. Yogi and Theta had two offspring that did not survive. (Photograph by Louise Cox.)

In 1983, Omaha Zoo sent Sol Yogi to the Knoxville Zoo on loan. Sol Yogi and Theta had a baby named Teddy in 1987. (Photograph by Jack Kirkland.)

This picture of the zoo's Mouflon was taken in April 1978. (Photograph by Rip Noel, *Knoxville Journal*.)

This handsome monkey is a cotton-top tamarin. They are so named because of the white ball of fluffy fur on the top of their heads that looks like cotton. Cotton-top tamarins were at the zoo from January 1971 until April 1989.

The Knoxville Zoo had South American tapirs from July 1976 until December 1991. Pictured here are tapirs Eggplant, Cucumber, and Bolivar (Bo) with keeper Geoff Gannicott. Bo came on loan from the Milwaukee Zoo in 1982.

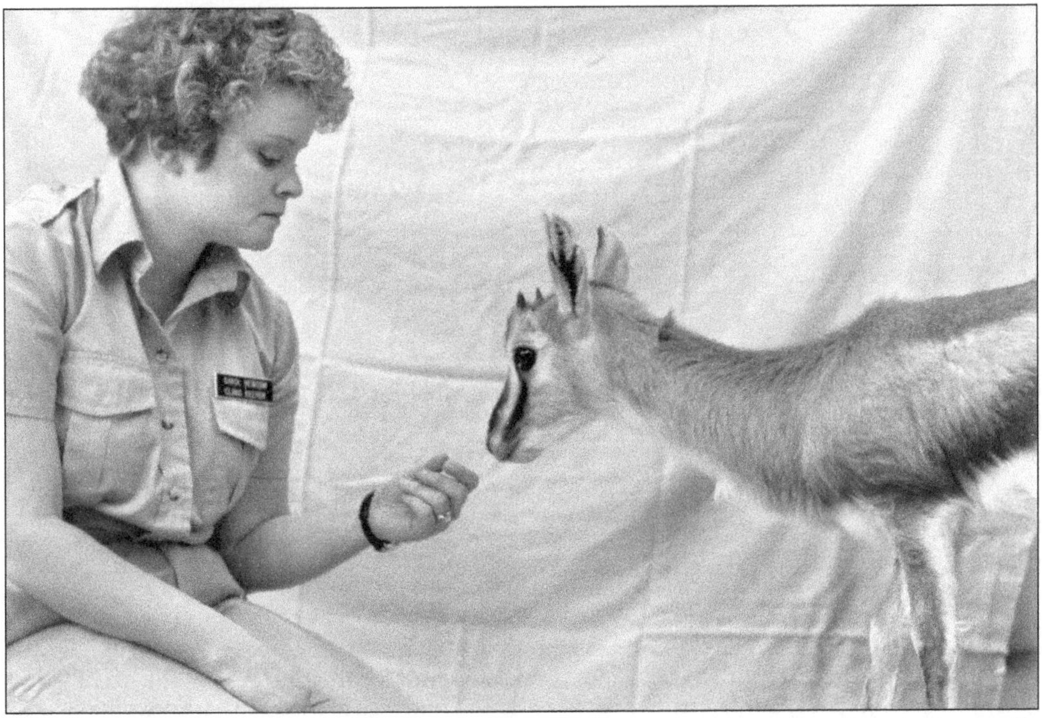

Tommy, pictured here with clinic technician Carol Newsom, was hand-raised at the Louisville Zoo. The Knoxville Zoo purchased the Thomson's gazelle from Louisville in December 1987.

These little Bengal tiger cubs were among the first litters of cubs born at the Knoxville Zoo.

Petrushka is pictured here with her two-month-old cubs in December 1978; Petrushka, a Siberian tiger, was born June 23, 1973, at the Lincoln Park Zoo and was loaned to the Knoxville Zoo in January 1975 for breeding purposes. Petrushka was later donated to the Knoxville Zoo, where she lived out her 16 years. (Photograph by Dave Carter, *Knoxville News-Sentinel*.)

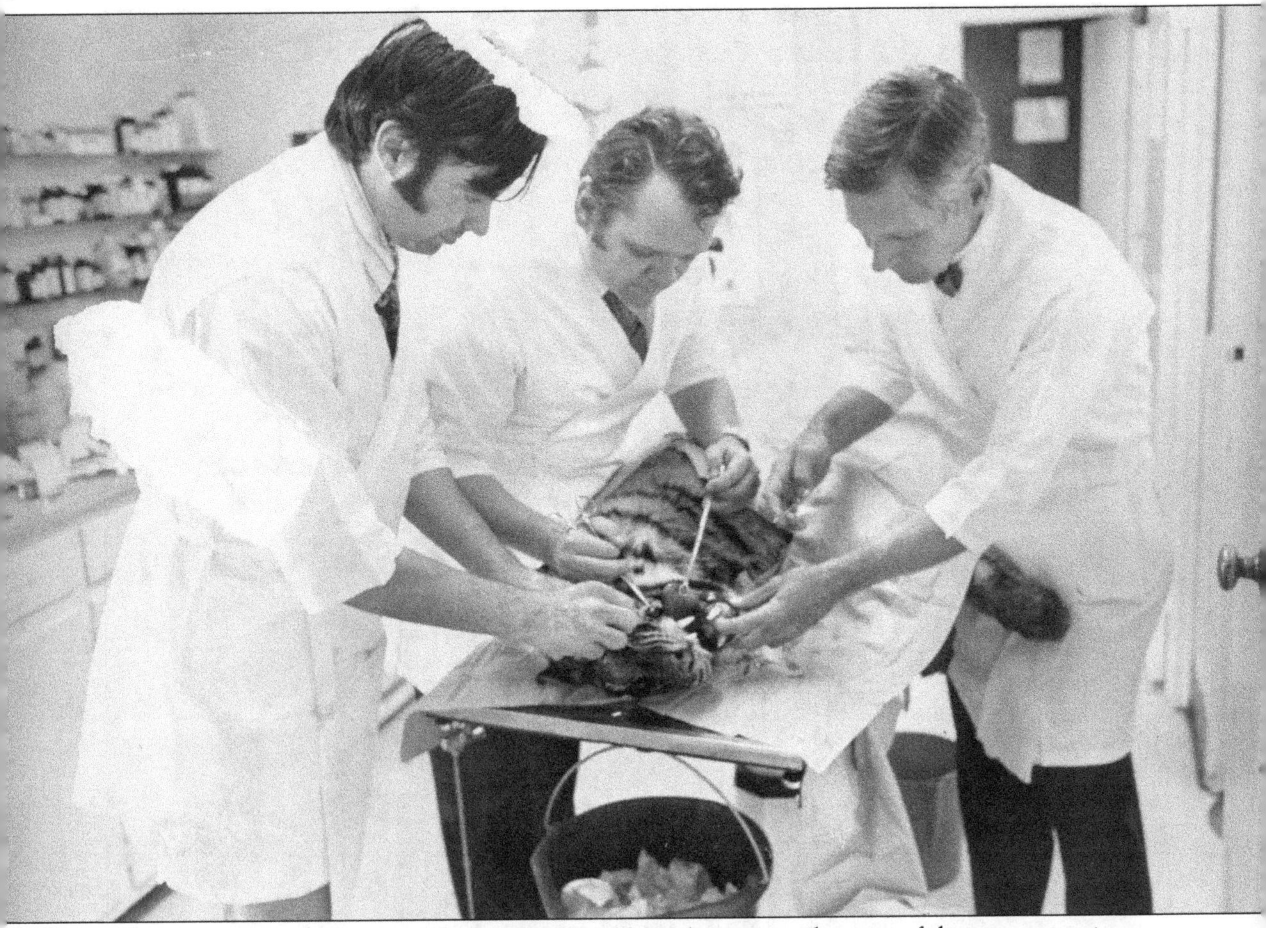

Vet visits are important for all zoo animals and have been a regular part of their zoo care since Guy Smith made an alliance with the University of Tennessee College of Veterinary Science in 1974. Prior to the arrangement with the University of Tennessee College of Veterinary Medicine, a local practitioner, Dr. William Montgomery, provided health care for zoo animals. Pictured from left to right are Montgomery (veterinarian), Dr. E. B. Andrews (plastic surgeon), and Dr. C. M. Hyder (oral surgeon) performing an exam on one of the zoo's Bengal tigers. (Photograph by Mickey Creager, *Knoxville News-Sentinel*.)

Sabu, a three-year-old Bengal tiger, came to the zoo from the Ringling Brothers, Barnum, and Bailey Circus in 1973. Sabu is shown here at the McGhee Tyson Airbase with the Air National Guard awaiting transfer to the zoo. (Photograph by *Knoxville News-Sentinel.*)

White Bengal tigers are very rare. All the white tigers in captivity in the United States today have descended from a common ancestor that was caught in India in 1951. Because the gene that causes the fur to be white is recessive, both parents must have the gene for the mutation to occur.

These yaks were photographed at the Knoxville Zoo in June 1975 by the *Knoxville News-Sentinel*. Yaks were exhibited at the zoo from 1975 through 1982. Like many species, yaks shown in zoos have been born in captivity and will most likely remain in captivity unless released to repopulate a particular area under the Species Survival Plan. (Photograph by *Knoxville News-Sentinel*.)

From 1976 to 2006, there have been 28 Southern white rhinoceroses born at the Knoxville Zoo, making it one of the nation's leaders in Southern white rhino births. Pictured here are rhinoceroses mom Tilly and her son Little Marvin in 1978. Fortunately the Knoxville Zoo has space to house several rhinos at once and can maintain more than one male, which may stimulate breeding. (Photograph by *Knoxville News-Sentinel.*)

Rhinoceroses are native to Africa and Asia, and while their large size makes them appear bulky and slow, they can actually charge at up to 35 miles per hour. In this 1978 photograph, Tilly teaches her young son Little Marvin to keep up. Though they are herbivores, rhinos can be dangerous and will charge when threatened. Rhinos can live to be over 50 years old. (Photograph by *Knoxville News-Sentinel.*)

In 1972, one could collect S&H Green Stamps to redeem for merchandise such as dishes, lamps, hair dryers, and, yes, even zebras. Local citizens saved their stamps for the city to redeem for zebras, and while there were some outspoken opponents to the acquisition, most people were willing to turn over their stamps to benefit the zoo. This photograph was taken shortly after they arrived. (Photograph by Hugh Lunsford, *Knoxville Journal*.)

When the city was collecting S&H Green Stamps to obtain two zebras, even congressmen and local schoolchildren participated. One little boy turned in a little green tag, hoping that his "stamp" would help purchase the animals. The zoo had to obtain two zebras at once because they are herd animals and would not do well alone.

WildCare

Adopt an Animal

KNOXVILLE ZOO

The grocery bill for the animals living at the Knoxville Zoo is almost $200,000 each year. The animals eat nearly 210,000 pounds of hay, 4,160 pounds of bananas, 2,500 pounds of meat, 500 heads of lettuce, and over 119,300 pounds of grain annually. Zoo visitors and others interested in the welfare of animals can adopt an animal through the WildCare program. The WildCare animal-adoption program makes a great gift and anyone can participate, including senior citizens, youth groups, Scout troops, classrooms, and individuals. For more information about adopting an animal and to find out what benefits this program includes, visit the Knoxville Zoo Web site at www.knoxville-zoo.org or call the zoo at (865) 637-5331.

Four

EDUCATION

Elephant keepers demonstrate one of many behaviors that elephants are trained to do when asked and that assist in their care and medical treatment. Most African elephants today are handled with protected contact rather than with the free contact shown here with Petunia.

In the spring of 1972, the docent program began at the zoo. The original docents were Dorothy Patterson, Carole Montgomery, Walter Boruff, and Suzi Hanna. The zoo has continued to maintain a docent program throughout the years and couldn't provide the exceptional educational programs without these volunteers. Shown here is Tim Adams, former education director at the Knoxville Zoo.

Bulletin

THE APPALACHIAN ZOOLOGICAL SOCIETY
BOX 283
KNOXVILLE, TENNESSEE 37901

Vol. 1, No. 1 January 1972
Larry L. Triplett, Editor

This is the first edition of the Appalachian Zoological Society Bulletin. Its goal is to keep you informed of the activities of the Society and the Zoo so that you may get involved and add your support in the Society's endeavors. The primary purpose will be to inform the general membership of the Society's current progress. In later editions, published every 6 to 8 weeks at first, we will attempt some detailed articles on such subjects as zoo animals, research projects at the zoo, goals of the Society, and other Society activities. If you have any suggestions concerning content, or format of the Bulletin, please let us know.

On December the 21st the Society had $750.00 in the bank with all bills paid. The Society had 108 members at the end of November (Individual-26, Family-57, Contributing Member-12, Patron-1, Benefactor-1, and Student-11).

OFFICERS OF THE SOCIETY
Mr. Guy Smith (Zoo Director)
Dr. W. L. Patterson (Chairman) Mr. R.W. Ritchie (Secretary)
Mr. Ben Hannah (V. Chairman) Mr. Steve Koella (Treasurer)
Mr. John T. Ellington (Membership Chairman)

Dr. Gordan M. Burghardt Mr. Lamar Mote
Mr. Gene Ford Dr. Mike Pelton
Mr. Jack Hanna Dr. Howard G. Senter
Mr. Julian Link

The Appalachian Zoological Society was formed in 1971. The society's goals were to improve conditions at the zoo, bring in new animals, and raise awareness in the community as to the benefits of a zoological park. Dr. William L. Patterson was the first chairman of the society. The society was instrumental in seeing many of the changes made that steered the zoo toward becoming the top-rate zoo that it is today.

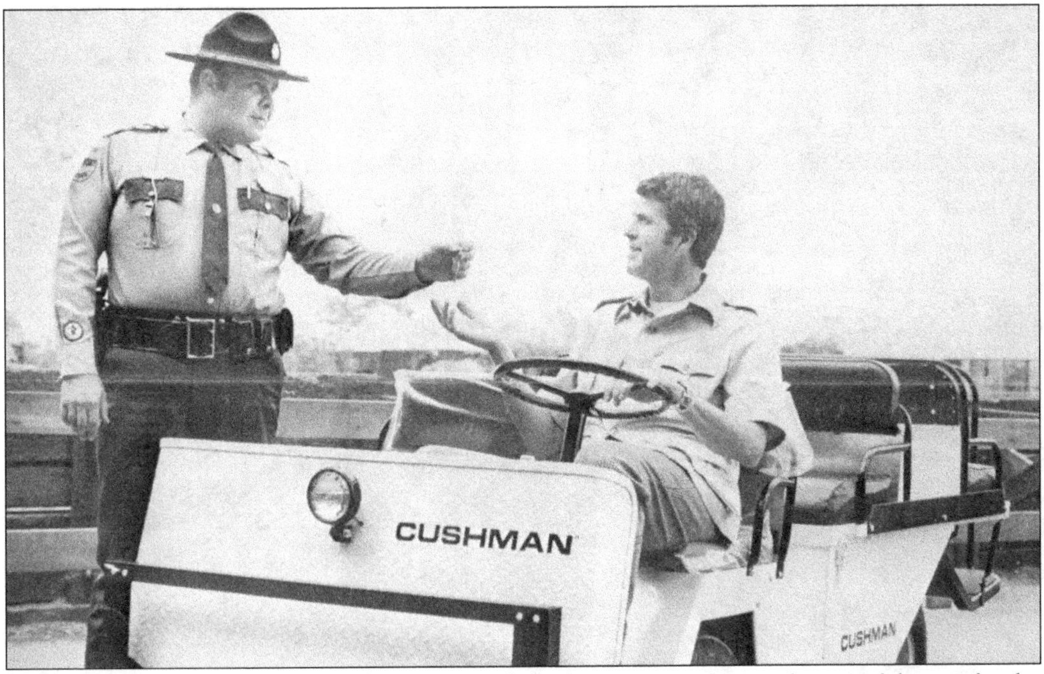

When Guy Smith took over the zoo, he not only had to improve the conditions of the zoo for the animals, but he wanted to improve conditions for visitors as well. With statistics showing that 30 percent of zoo visitors were handicapped, in 1976, Tennessee State Troopers donated two large golf carts for docent-led education programs and to transport handicapped individuals around the zoo. Today the zoo is accessible for those with wheelchairs and strollers.

Birds have never been exempt from Encounters (docent-led educational programs with the animals), and this early photograph shows bird keeper Lynn McCoig preparing to do an educational program with a parrot. (Photograph by Arthur J. Miller Jr.)

Elephants are always favorite animals at any zoo, and the Knoxville Zoo is no different. Education programs about elephants might include what they eat, how long they live, where their native lands are, how dangerous they are, and how each of them has different personalities. In these two photographs, the elephants perform during Encounters and prove to the audience that they truly are quite intelligent animals.

The Dorothy W. and William L. Patterson Log Cabin was dedicated on May 22, 1976, with, from left to right, Dr. William Patterson; Dorothy Patterson; Bill Arant, board chair; and Wally Estill, docent council chair. The log cabin serves as an educational building and as the location of zoo birthday parties.

Zoo keeper Greg Sewell allows a young boy to touch an alligator. Many of the animals children (and adults) encounter at the zoo are not found commonly in residential areas. Going to the zoo and seeing animals, particularly attending educational programs and actually seeing the animals up close and touching them, allows people to develop a fondness for these creatures, which need our protection.

Nancy Summers, a docent at the Knoxville Zoo, explains biofacts to a group of interested children. Docents educate visitors about all sorts of interesting animal facts, such as how much they eat, what they eat, how long they sleep, and their typical life spans. Education has always been a priority at the zoo, and visitors crowd around when the docent-led presentations, now called Encounters, takes place.

Visitors are very interested in this snake being shown by Tim Adams at a zoo Encounter.

Jack Davis has been an employee at the Knoxville Zoo since 1997. The Otter Jack's souvenir stand was named for him, and perhaps an even greater honor was bestowed when Marie Vlna, director of guest services, created "You Too Can Do the Zoo." The Knoxville Zoo has been aggressively implementing changes to walkways and exhibits throughout the zoo in order to make the grounds as accessible as possible to all visitors. When trying to create a special day for visitors with special needs, Marie had Jack in mind. Jack has cerebral palsy and must use a wheelchair. Marie and other zoo staff wanted to create a day when individuals such as Jack could enjoy the day for free and also have the assistance of volunteer interpreters if needed.

The naked mole rats exhibit turned out to be extremely popular with zoo visitors. It is actually a traveling exhibit, and that changes periodically. The building that houses the exhibit is the Pilot Traveling Exhibits Building and it is sponsored by the Haslam Family Foundation/Pilot Corporation.

The naked mole rat gets its name from the fact that it has almost no hair. The rats are burrowing rodents and live underground. The tunnels they dig can extend for miles. Each rat in a mole rat colony has a specific job, whether that job is to dig tunnels, move debris from the tunnel, care for the baby mole rats, or serve as soldiers to protect the colony from enemies.

Bedtime with the Beast instructors Evan and Teri James lead a group of blindfolded children through the zoo to prepare for the overnight educational program. Evan began volunteering at the zoo at age 13 and then became a zoo employee at age 15. He is currently curator of Kids Cove. Bedtime with the Beasts provides children the opportunity to interact with particular animals and tour the zoo in a nocturnal setting. The program started in 1993 with Paula Evans as coordinator.

During Zoo Lights Before Christmas, zoo visitors took evening strolls to view thousands of Christmas lights, drink hot chocolate, eat roasted chestnuts, hear stories from Mrs. Claus, and see certain animal exhibits. The event ran from 1989 to 1993, when it was discontinued because it was very labor intensive and income did not even cover expenses. This photograph was taken during Zoo Lights in 1989.

Five

Activities and Special Events

During the Zoo Olympics from the late 1980s to early 1990s, guests formed teams and participated in various events such as racing, tug-of-war with the elephants, and wheelbarrow races. In this photograph, guests stack blocks during the 1993 Zoo Olympics.

Penny Beattie, Barbara Arant, and other members of the Zoofari committee participate in an Elephant Wash in late 1988. Zoofari is an annual fund-raising event that began in 1988 and still exists today. Each year, the Zoofari Committee plans a theme and the events to go with it. There is normally a cocktail party, a live auction, and a seated dinner with a live band. This annual event currently raises over $100,000 for the zoo.

The Teddy Bear Picnic was a popular special event at the zoo in the 1980s. This was during the era when Ted Beattie was doing a lot to promote the zoo in the community and raise awareness of all it has to offer. Like many large events at the zoo, they rely on volunteers to help make these activities a success. This photograph shows Dr. Mike Shires, aka "Dr. Fuzzy-Wuzzy," dean of the University of Tennessee College of Veterinary Medicine.

Here a young girl checks out her teddy bear with a stethoscope. Many volunteers came to the zoo to participate in the Teddy Bear Picnic. These volunteers bandaged "wounded" or torn bears, helped little children do exams on their bears, and judged bears in the bear contests—biggest, oldest, fluffiest, and others. The Teddy Bear Picnic was sponsored by Thompson Photo Products.

Easter Eggcitement began on March 25, 1989, as a zoo-alternative Easter event for Easter egg hunts. Zoo volunteers and staff donated 140 dozen eggs for the event, and there were 165 participants that first year. As you can tell from this photograph, the Easter bunny came to greet guests, then they participated in egg decorating, an Encounter with a live bunny, and an egg hunt.

In 1972, a Concert in the Park began as a fund-raiser. Pictured here, Mark and Kathy McGranaghan enjoy the music with their children (from left to right), Ryan, Maureen, and Andrew.

Boo! at the Zoo is Knoxville's largest Halloween event, drawing over 30,000 guests the two weekends before Halloween. This is a fund-raiser for the zoo, and most animals are not viewable during this time since this is usually when they sleep. Though some of the characters appear scary in this photograph, this is no longer the case with Boo! at the Zoo.

The Municipal Zoo had no admission fee, and dilapidated conditions reflected this lack of funding. To improve conditions, in the spring of 1972, the zoo began charging an entry fee of 50¢ for adults and 25¢ for children. In 1974, the zoo hosted a Free Zoo Day, when patrons could visit at no cost. Unfortunately, guests did thousands of dollars worth of damage to shrubbery, plants, and flowers, and Guy Smith, zoo director, stated that it would be the last such "free" day. However, the zoo later decided to resume the treat for the community, and pictured here are visitors streaming past what is now the Stokely African Elephant Preserve area on a later Free Zoo Day.

Thompson Photo Products was founded in 1902 and operates today as a fourth-generation business. In this photograph, Thompson Photo sponsored a photograph day when zoo visitors could use cameras on loan from Thompson's to take pictures of the animals. The visitors then took their rolls of film to be developed. It was a good reciprocal sponsorship because the photograph store received exposure and the zoo's animals received further appreciation.

The Knoxville Zoo sponsored a float in a Knoxville parade in 1998. Zoo staff and volunteers dressed up like animals or carried small stuffed animals to show observers.

Six

Children's Area

Several young girls visit animals in the children's petting zoo at the Municipal Zoo in the early 1970s. The petting area is always a popular place for children, and the one at Knoxville Zoological Gardens is no different.

Camels are a popular animal among children and adults. The humps on their backs, their ability to go without water for long periods of time, and their friendly looking faces seem to attract young and old alike. Here young visitors to the Knoxville Zoological Park enjoy a camel ride. (Photograph by Jessica Stooksbury.)

Boy Scout Michael Porter pays a visit to the deer in this early portrait at the children's petting-zoo area. The petting area has always been a popular spot for children of all ages. To ensure the safety of each child who visits the petting area, the animals in it are selected for their docile manners, easy handling, and small sizes.

Melissa Gray enjoys a ride on the back of Little Diamond in May 1981. Trainer Frank Huskey stands nearby. Melissa's mother, Linda Gray, later began volunteering at the zoo as a docent. She continues to work there doing Encounters for zoo visitors. (Photograph by Linda Gray.)

A young girl sits on Petunia during the ZooFest of 1978.

The zoo would not be able to exist without volunteers. Numerous volunteers come in on a regular basis and even more come in to help with special events. B. Grater was a petting-zoo volunteer for numerous years, and here in 1984, B. speaks with a young visitor. (Photograph by Betty Wasserman.)

Maintaining beautiful grounds has always been a priority for the Knoxville Zoo. In more recent years, however, zoo designers have placed a growing priority on having particular species of trees, grasses, plants, and flowers throughout the zoo that complement the animals living there. This view is near the old entrance looking back toward what is now Kids Cove. (Photograph by Keith Montgomery.)

Zoo volunteer B. Grater visits with children and goats in the zoo's petting area in 1987. (Photograph by Betty Wasserman.)

Jason Stooksbury sits on the ground in the children's petting-zoo area while a goat considers munching his hair. Goats are known to chew on anything loose, so visitors are encouraged to keep a diligent eye when they are in the petting-zoo area. The goats might just think a little girl's hair or a little boy's loose shirt looks as good as a clump of grass. (Photograph by Betty Wasserman.)

Animal keeper Pat Parks helps a young visitor get a look at the new Sicilian donkey standing beside her mother in the petting zoo. Mary Jane was born October 9, 1985, and is pictured here with her mother, Suzannah, just a few weeks after her birth. (Photograph by Betty Wasserman.)

The Municipal Zoo was able to add Big Boy to the elephant herd in November 1972 thanks to a generous donation from Ray Staab (right), vice president and general manager of Shoney's restaurant. Of course, the elephant was named in honor of the restaurant mascot. Staab is pictured here with Dr. William Patterson, chairman of the Appalachian Zoological Society. The society was created by Guy Smith, Jack Hanna, Dr. William Patterson, and a few other individuals in 1971 to run operations at the zoo. Big Boy was purchased from Frank Thompson and was estimated to be about two years old when he came to the zoo.

The narrow-gauge railroad began operating in April 1985 with a live steam engine and six passenger cars. The 15-inch-gauge engine was built at Janesville, Wisconsin, in 1948 by Sandley Locomotive Manufacturing Company. It came to Knoxville for the 1982 World's Fair, then went to the Knoxville Zoo. (Photograph by Ron Warwick Studio.)

After coming to the Knoxville Zoo in 1985, the narrow-gauge steam engine was eventually replaced with a more powerful diesel engine built in 1988. The zoo train ran until September 1995, when it was dismantled because it needed repairs on the half mile of track and 400-foot-long trestle that the zoo couldn't afford to make. The train equipment was sold to the Railstar Corporation in New York.

This staged photograph was used as a Knoxville phone-book cover in the mid-1980s. Normally elephants, tortoises, and peafowl didn't run around together near the old entrance area in front of the grottos. The objective of the photograph appears to have been simply to get as many different types of animals in the picture as possible. (Photograph by Bill Tracy.)

Petunia prepares to lead the grand parade at the Teddy Bear Picnic in July 1988. The zoo has always sponsored many activities for children, some of which can be seen in the last chapter. Since children will continue to be the individuals who care for our animals in the future, the zoo will continue to educate them and demonstrate to them about various animals at the zoo. (Photograph by Betty Wasserman.)

One of the benefits of becoming a zoo member is being able to attend members-only events. This can sometimes mean shorter lines and less crowding at the most popular exhibits, but it always means a great deal for your family and the zoo. In this picture, members enjoy a special Members' Night on August 28, 1988. (Photograph by Betty Wasserman.)

Kids Cove, sponsored by the Clayton family, opened April 2, 2005. Children can climb through a log cabin, play in a sandbox, climb on a "spider web" made of rope, play with animals in the contact yard, slide down twisting tubes, see beavers chew wood and make a dam, watch birds in an aviary, and look at eggs in a chicken coop. There is also a water area where children are encouraged to play on lily pads and where giant frogs that shoot water from their mouths. Kids Cove is meant to attract children ages 1 to 10, and there is something to do in this area year-round. There are also frequent educational programs.

The top photograph was taken behind the old elephant area. If you were standing in the same place today, you would see the carousel that was specially designed for the Knoxville Zoo.

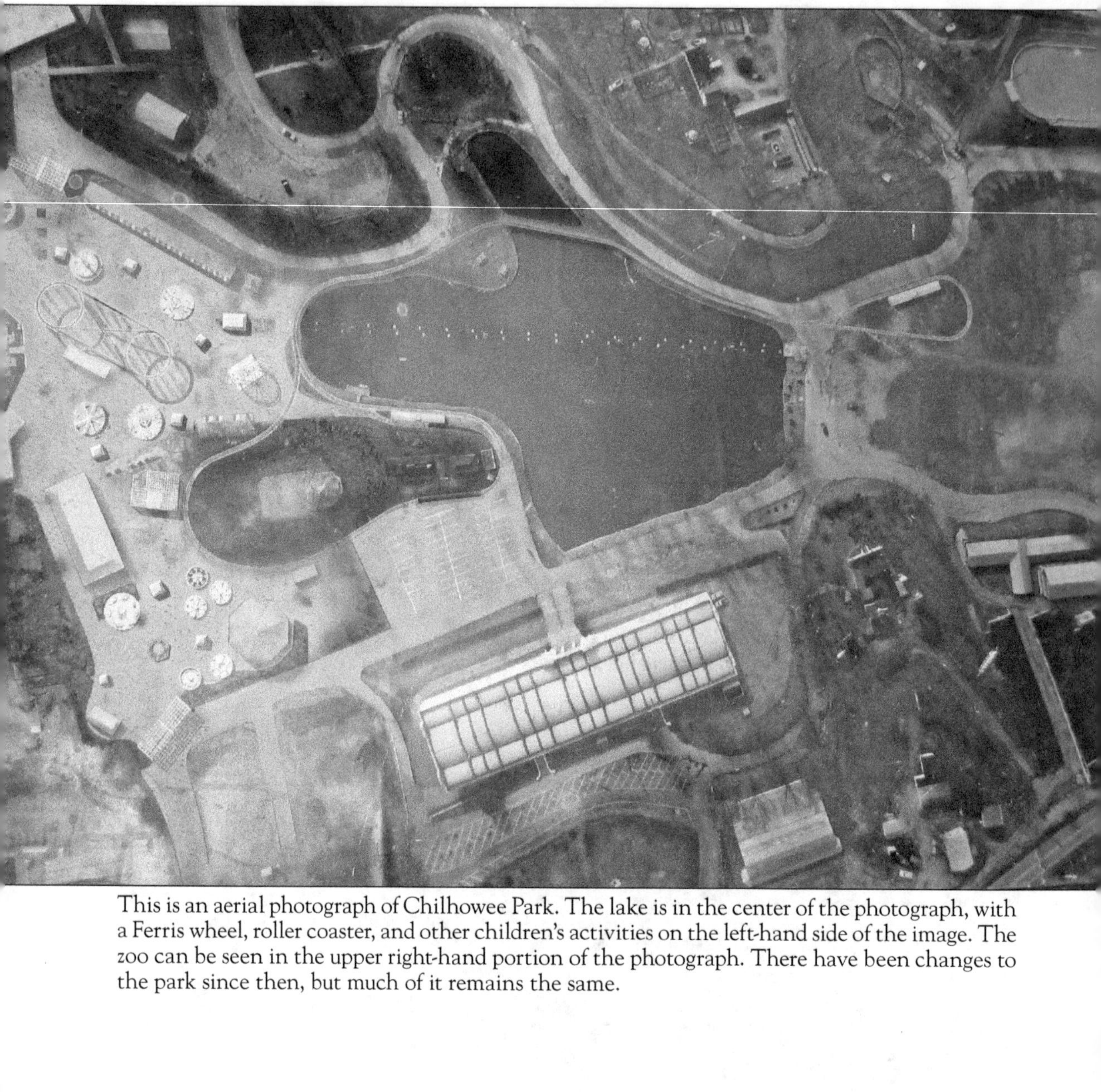

This is an aerial photograph of Chilhowee Park. The lake is in the center of the photograph, with a Ferris wheel, roller coaster, and other children's activities on the left-hand side of the image. The zoo can be seen in the upper right-hand portion of the photograph. There have been changes to the park since then, but much of it remains the same.

BOARD MEMBERS AND PAST CHAIRMEN

Knoxville Zoological Gardens, Inc.
Board of Directors 2006

OFFICERS
Chairman of the Board: David Moon
Vice Chairman: H. Pete Claussen
Secretary/Treasurer: Will J. Pugh Sr.

Dr. Michael Blackwell
Mark Brown
David Collins
Cynthia Finch
Sam Furrow (past chairman)
Mack Gentry
Dee Haslam
J. E. Henry
Christopher Kinney
Jeff Lee
Bob Polk
Patrick Roddy
Robert S. Talbott
Tim Williams
Robert F. Worthington Jr.

HONORARY DIRECTORS (PAST CHAIRMEN)
Dr. William L. Patterson
James A. Bean
Williams E. Arant Jr.
William W. Baxter
Lee Congleton
Roger Daley
Justice Gary Wade

ACKNOWLEDGMENTS

This book would not have been possible without the support and encouragement of several people who supported the project wholeheartedly from the beginning. Most of all, I would like to thank Jim and Marie Vlna for giving me the opportunity to compile this book and share the history of this beautiful zoo. Marie also served as a valuable resource for information, always answering my questions, providing information, and cheerfully referring me to others when needed.

Several zoo staff provided assistance identifying dates and subjects, whether the subjects were people or animals. A special thank-you goes to Jane Kading, Linda James, and Keith Montgomery for being available to help.

Finally, I would like to thank my family for being patient and supportive while I worked. I would not have been able to use the gifts God has given me if it weren't for my husband, Chris. He is forever supportive of all my projects and activities. Also my children compel me to write books that are not only appealing to historians, librarians, tourists, and others, but books that are also pleasurable and appropriate for young people. I thank Hannah, Daniel, Christopher, Micah, and Sarah for making every day of my life memorable.

There are many other individuals and groups who have helped make the zoo what it is. Space only allows mention of a few of these people, but the zoo appreciates everyone who works to preserve the animals of our world. On behalf of the zoo, I send a special thank-you to City of Knoxville mayor Bill Haslam, the Knoxville City Council, Knox County mayor Mike Ragsdale, and the Knox County Commission for their continued support of the Knoxville Zoological Gardens.

About the Author

Sonya A. Haskins has a bachelor's degree in English from King College in Bristol, Tennessee. She has written hundreds of articles for newspapers and magazines, including *Writer's Digest*, *Guideposts for Teens*, *Decision*, *Physician's Practice Digest*, *Educational Dealers*, and *Homeschooling Today*. Sonya has written several books for Arcadia Publishing, including *Johnson City*, *Jonesborough*, *Bristol*, and *Soddy-Daisy* as part of the Images of America series. She coauthored *Bristol Motor Speedway*, in Arcadia's Images of Sports series, with David M. McGee. All the books have been local best-sellers. Sonya is very active as a local home educator. She enjoys meeting and providing support for parents interested in teaching their own children at home. Sonya is a motivational speaker and welcomes opportunities to do presentations at schools, nonprofit organizations, home education support groups, and Bible studies. She is available via e-mail at sonya@sonyahaskins.com. (Photograph by Sarah E. Haskins.)

www.ingramcontent.com/pod-product-compliance
Lightning Source LLC
LaVergne TN
LVHW081933200926
848484LV00009B/2414